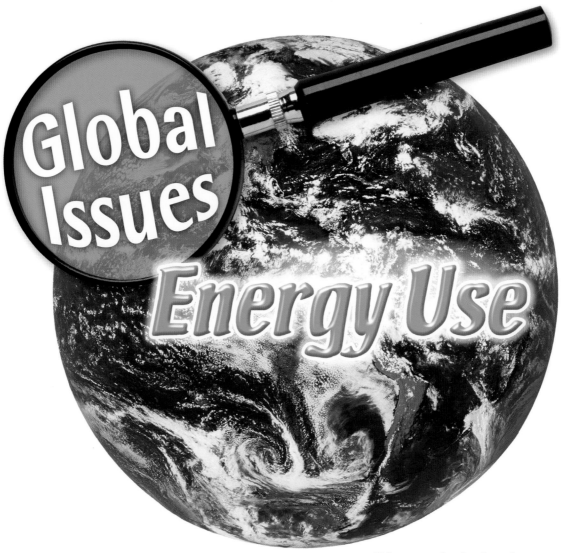

Global Issues

Energy Use

Cheryl Jakab

Smart Apple Media

This edition first published in 2008 in the United States of America by Smart Apple Media.

Smart Apple Media
2140 Howard Drive West
North Mankato, Minnesota 56003

First published in 2007 by
MACMILLAN EDUCATION AUSTRALIA PTY LTD
627 Chapel Street, South Yarra, Australia 3141

Visit our Web site at www.macmillan.com.au or go directly to www.macmillanlibrary.com.au

Associated companies and representatives throughout the world.

Library of Congress Cataloging-in-Publication Data

Jakab, Cheryl.
    Energy use / by Cheryl Jakab.
    p. cm. — (Global issues)
    Includes index.
    ISBN 978-1-59920-126-9
    1. Power resources. 2. Energy consumption. I. Title.

    TJ163.2.J33 2007
    333.7913—dc22

                                                    2007004562

Edited by Anna Fern
Text and cover design by Cristina Neri, Canary Graphic Design
Page layout by Domenic Lauricella and Cristina Neri
Photo research by Legend Images
Illustrations by Andrew Louey; graphs by Raul Diche
Maps courtesy of Geo Atlas

Printed in U.S.

Acknowledgements
The author and the publisher are grateful to the following for permission to reproduce copyright material:

Front cover inset photograph: Looking across New York City at night, © Stephen Price/Fotolia. Earth photograph courtesy of Photodisc.

Background photograph of Earth and magnifying glass image both courtesy of Photodisc.

DOE/NREL/Chris Gunn Photography, p. 11; DOE/NREL/Glickson, Amy-Pixl Studio, p. 26 (left); © Bmcent1/ Dreamstime.com, p. 5; © Brm1949/Dreamstime.com, p. 20; © Carolinasmith/Dreamstime.com, p. 22; © Ib2loud/ Dreamstime.com, p. 9 (top); © Urbanraven/Dreamstime.com, p. 15; © Ian Holland/Fotolia, p. 12; © William Mckelvie/Fotolia, p. 14; © Stephen Price/Fotolia, pp. 6 (left), 8 (bottom); © Pete Fleming/iStockphoto.com, p. 23; ITER, published with permission, p. 27; Photolibrary, p. 7 (bottom), 21; Photolibrary/Science Photolibrary, pp. 18, 24; Rob Cruse Photography, p. 10; © b.p. khoo/Shutterstock, pp. 7 (right), 17; © Timothy Large/Shutterstock, p. 19; © Albert Lozano/Shutterstock, p. 29; © Dianne Maire/Shutterstock, pp. 6 (top), 25; US Navy/Kevin H. Tierney, pp. 6 (bottom), 13.

Please note
At the time of printing, the Internet addresses appearing in this book were correct. Owing to the dynamic nature of the Internet, however, we cannot guarantee that all these addresses will remain correct.

Contents

Glossary words
When a word is printed in **bold**, you can look up its meaning in the glossary on page 31.

Facing global issues

Hi there! This is Earth speaking. Will you take a moment to listen to me? I have some very important things to discuss.

We must face up to some urgent environmental problems! All living things depend on my environment, but the way you humans are living at the moment, I will not be able to keep looking after you.

The issues I am worried about are:
- the huge number of people on Earth
- the supply of clean air and water
- wasting resources
- energy supplies for the future
- protecting all living things
- **global warming** and **climate change**

My global challenge to you is to find a **sustainable** way of living. Read on to find out what people around the world are doing to try to help.

Fast fact

In 2005, the **United Nations Environment Program Report**, written by experts from 95 countries, concluded that 60 percent of Earth's resources are being **degraded** or used unsustainably.

What's the issue?
Energy for the future

Across the world today, oil, coal, and gas are the main sources of energy for industry, transportation, and the home. Oil, coal, and gas are **fossil fuels**. Developing energy sources as alternatives to these fuels is an urgent environmental issue.

The need for energy

Everything that happens needs an energy source. Human bodies use food for energy and plants use the energy of sunlight to grow. Since the 1850s, people have come to rely more and more on fossil fuels as their main energy sources, and most modern machines still use oil or coal.

Problems with fossil fuels

The are two main problems with relying on fossil fuels:
- fossil fuels are **nonrenewable** energy sources, and they will run out
- using fossil fuels causes environmental problems, including pollution and global warming

Overcoming these problems requires the development of a range of alternative fuel sources.

Fast fact
Until the mid-1800s, the main fuel source used by people was wood.

Most of the energy used by people today comes from fossil fuels.

GASOLINE

Energy issues

The most urgent energy supply issues around the globe include:
- the high demand for energy (see issue 1)
- supplies of nonrenewable fossil fuels rapidly running out (see issue 2)
- pollution and global warming caused by burning fossil fuels (see issue 3)
- running out of **renewable** fuels such as wood (see issue 4)
- problems with new alternative fuel (see issue 5)

Arctic Circle

NORTH

AMERICA
United States

NORTH

ATLANTIC

OCEAN

SOUTH

AMERICA

SO

ATLA

OC

ISSUE 1

United States
The highest energy user.
See pages 8–11.

ISSUE 2

Saudi Arabia
Even the largest oil reserves have a limit. See pages 12–15.

around the globe

Fast fact
Fossil fuels form under the ground from the remains of plants and animals that lived millions of years ago.

ISSUE 5

France
Turning to the nuclear energy alternative. See pages 24–27.

Arctic Circle

EUROPE

France

ASIA

China

Saudi Arabia

AFRICA

Equator

ISSUE 3

China
Air pollution due to high use of coal. See pages 16–19.

Tropic of Capricorn

AUSTRALIA

TIC

N

ANTARCTICA

ISSUE 4

Africa
Problems with wood as a source of energy. See pages 20–23.

High energy demand

Modern **developed countries** have very high energy demands. They need huge amounts of energy every day to run machines in manufacturing, in offices and homes, for communications, and for transportation.

World energy demand

The world population is rapidly increasing and more and more people across the globe are using more and more energy each year. Since the 1850s, industrial development has depended on fossil fuels for energy. Oil, coal, and gas are still the main energy sources used today.

Fuel consumption today

Fuel consumption worldwide is increasing fast. Nations with well developed industry are continuing to demand more and more energy to run more and more devices. Many people in **developing countries**, particularly in Asia, Africa, and South America, are moving from a rural lifestyle to an industrial lifestyle with a high energy demand.

Fast fact

In 2000, world consumption of fossil fuels was about 30 times greater than it was in 1900.

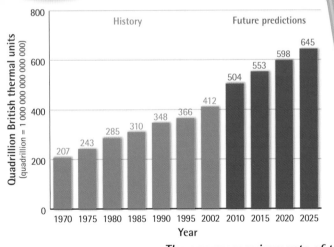

The energy requirements of the world population continue to rise.

Electricity has many uses in the modern world.

The U.S. is rich in fossil fuel resources such as oil.

Fast fact
In worldwide fossil fuel reserves, the U.S. has more coal than any other country. It has the sixth largest natural gas reserves, and the eleventh largest oil reserves in the world.

CASE STUDY

The U.S.—world's largest energy consumer

Of all the countries in the world, the U.S. consumes by far the most energy. The U.S. is the world's largest energy producer, consumer, and importer.

It is predicted energy consumption in the U.S. will increase by 25 percent by the year 2025. Transportation will account for most of this growth in demand for oil through to 2020.

Cars in the U.S.

Cars account for a large part of the huge demand for oil in the U.S. About 80 percent of total travel each year is by private car. Airline travel is the second most popular way to travel. Very little travel is by public transportation such as buses and trains. Passenger vehicles in the early 2000s used about 40 percent of all the petroleum consumed in the U.S.

ISSUE 1

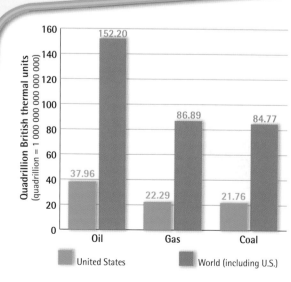

The U.S. has about 5 percent of the world population and consumes about 25 percent of the energy.

9

Toward a sustainable future: Reducing consumption

In the early 2000s, about 20 percent of the world's population in developed countries consumed about 80 percent of the world's resources, including energy. Reducing energy consumption can save large amounts of energy resources so they will last longer.

Energy efficiency

Energy efficiency is the quickest and cheapest way to extend the world's energy supplies. Efficient devices use less energy to do the same job. Energy-efficient lightbulbs, cars, washing machines, and refrigerators all help save energy supplies by reducing the energy needed to do a job.

Switch off and save

Switching off appliances when they are not in use decreases energy demand and saves money. Electric lights left on, electrical appliances on standby, and using cars to take short trips instead of walking all waste energy.

Fast fact
It is estimated that using energy-efficient household appliances and improved technologies can result in a 50 percent saving in electricity.

Energy-efficient appliances help reduce energy consumption.

When tilted down, these window coverings stop direct heat from the sun. When tilted up, they act as light scoops to reflect light into the living area.

CASE STUDY

Passive solar building design

Buildings can be designed to use the sun for heating or cooling, rather than using air conditioners and heaters. This use of the sun's energy is called "passive solar energy." Buildings can also be designed to make the most of natural lighting from the sun during the day.

Passive solar features

Passive solar designs include:

- windows that collect extra heat in winter and less heat in summer
- air channels to direct heat into the house in winter and keep heat out in summer
- thick walls that prevent swings in temperature by absorbing heat in winter and insulating in summer

Advantages of passive solar design

Designing buildings to better use passive solar energy can make huge energy savings. Passive solar design can dramatically reduce the amount of energy used and therefore save money on energy bills.

Limited oil supplies

Today, oil is used about four times as fast as new supplies of oil are found. At this rate, the known oil reserves could be used up in as little as 45 years.

What is oil?

Crude oil or petroleum is a naturally occurring, oily liquid which forms from long-dead plants and animals. Most oil formed below the oceans from the remains of tiny animals.

Crude oil is the main fuel for transportation. Oil is also used in medicines, fertilizers, foods, plastics, building materials, paints, cloth, and to generate electricity. **Lubricants** made from petroleum are used in just about every machine.

Reserves of crude oil

Oil is not distributed evenly through Earth's crust. Nearly 70 percent of the world's known oil reserves are in the Middle East, in Saudi Arabia, Iraq, Iran, the United Arab Emirates, and Kuwait. About half of the crude oil taken from these reserves is refined for use as gasoline in cars and trucks.

Fast fact
Producing and transporting oil often results in oil spills, which can have disastrous effects on the environment.

Offshore wells are the source of about a quarter of the world's annual production of oil.

Oil is transported in supertankers.

CASE STUDY
Oil in Saudi Arabia

Fast fact
One barrel contains 42 gallons (159 l) of oil.

Saudi Arabia produces about one-eighth of the world's oil each year. Production was estimated in 2001 to be over 3 billion barrels.

Saudi Arabia's oil reserve is not accurately known, although it is estimated at more than 250 billion barrels. It is difficult to know exactly, but at current rates of production, oil in Saudi Arabia may last for 80 years at the most.

Oil production and export

Most oil produced by Saudi Arabia is exported through ports in the Persian Gulf. The Trans-Arabian Pipeline, completed in 1950, carries oil to Lebanon. Another pipeline completed in the 1980s links the eastern oilfields with the Red Sea.

Oil pollution

Saudi Arabian **habitats** experience a number of problems from oil production. Oil wells, pipelines, and ships create oil spills on land and in the sea, causing extensive pollution problems.

Toward a sustainable future: Ensuring future energy supplies

Oil is a nonrenewable resource and will not be the main fuel used for very far into the future. The future of energy supplies can be ensured by:

- using oil reserves carefully and wisely
- finding alternatives to oil

Using reserves wisely

Remaining oil reserves need to be conserved now to make sure supplies last as long as possible. Experts agree that new discoveries and inventions can extend the availability of cheap oil for only a few decades. New technologies may also allow more oil to be taken from known deposits.

Wind turbines, which spin in the wind and make electricity, are being developed as a renewable energy source.

Finding alternatives

Renewable alternative energy sources include wood, wind, the sun, energy from ocean waves, and **hydroelectricity**. The best alternative sources to use are ones that:

- cause little environmental damage
- give a continuous supply that will not run out

Fast fact

In October 1994, an estimated 60,000 to 80,000 tons of oil were spilled from a fractured pipeline near Usinsk, just south of the Arctic Circle.

The Alaskan oil pipeline carries oil from oilfields on the Arctic coast to tanker ships.

CASE STUDY
New oil deposits

The search continues for new oil deposits to help keep up the supply. Additional discoveries of new oilfields will be made, but there are limits.

Oil in Alaska

The Prudhoe Bay oilfield on the North Slope of Alaska is the largest oilfield ever discovered in the Americas. The crude oil in this field is estimated to be about 10 billion barrels.

The Alaskan oil pipeline brings crude oil from the Prudhoe Bay oilfield to tanker ships docked in southern Alaska. Crossing 800 miles (1,290 km) of Alaskan wilderness, the pipeline carries up to 2 million barrels of oil per day from the Arctic coast to the Gulf of Alaska.

At current rates of oil use, Prudhoe Bay will only produce enough oil to supply the United States for less than two years. Prudhoe Bay is the only oilfield of this size discovered in the area in more than 100 years of exploration.

Fast fact
It is estimated the remaining oil reserves in the U.S. will last for less than 10 years at the current rate of production.

Pollution from burning coal

Coal is a major energy source, used particularly for heating and generating electricity in power stations. A major disadvantage of coal, and to a less extent other fossil fuels, is the release of large amounts of pollution into the air when it burns.

What is coal?

Coal is a solid fossil fuel formed from plant material. It consists mainly of carbon, a material found in living things. Most known coal reserves formed between 345 and 280 million years ago.

Coal is a rich energy source, but not as rich as oil. It is also not as easy to get from the ground or transport as oil.

Pollution from coal

Use of coal and other fossil fuels for energy is thought to be the main cause of global warming being experienced today. Burning coal releases:

- **carbon dioxide** that adds to global warming
- sulfur dioxide, which causes **acid rain**
- fine particles of air pollutants causing serious health problems

Coal forms under the ground over millions of years.

Step 1

Prehistoric plants grow on the land.

Step 2

Dead plants are covered as sea levels rise.

Step 3

Coal

Plant material is pressed into coal under the sea.

Step 4

Coal

The sea level drops, uncovering land with coal deposits.

Burning coal is the major source of air pollution in China.

CASE STUDY
Coal use in China

Coal is the leading industrial and domestic fuel in China. Coal is used to generate most of China's electricity and is causing enormous pollution problems. China's coal-mining industry is the world's largest.

Coal reserves in China

China has more than 40 percent of the world's known reserves of coal. Chinese coalfields are among the largest in the world, with over 10 trillion tons of coal, mostly in north China, around Dongbei. There are also many smaller coal mines throughout the country.

Air quality and coal

Using its huge coal reserves has helped China develop rapidly as an industrial nation. Like many similar places around the world, the air quality of industrial areas in China suffers greatly from air pollution.

Fast fact

Coal provides about three-quarters of China's energy. China also generates a significant amount of electricity by flowing water or hydroelectric power.

Toward a sustainable future: Finding new fuels

Coal is limited in supply, and reserves of coal should last about another 200 years if used at the 1998 levels of production. However, the biggest problem with coal is the pollution it causes when it burns. Alternative renewable fuel sources are needed to replace nonrenewable, polluting fossil fuels.

Alternative energy sources

Renewable energy sources such as wind, wave, and solar energy, are being developed to replace coal for generating electricity. Nonrenewable alternatives in use include **geothermal** energy and nuclear energy.

Clean coal technologies

New technologies for using coal are also being investigated to try to reduce pollution. These "clean coal technologies" reduce pollution given off by burning coal, and also get more usable energy from the coal.

Clean coal technologies include:
* improved methods of cleaning coal before use
* removing pollutants from wastes instead of releasing them into the air
* using wastes as a fuel source
* carbon sequestration, which is burying waste carbon dioxide deep underground

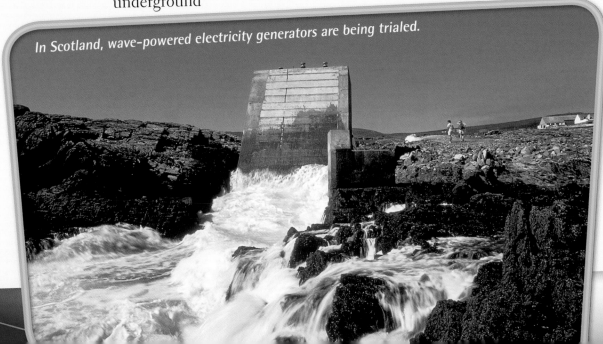

In Scotland, wave-powered electricity generators are being trialed.

Solar panels can be installed in individual buildings.

CASE STUDY
Solar energy

A major alternative to coal as a source of electricity is solar energy. Solar energy is used in large-scale power plants and on a small scale in individual homes.

The world's largest solar power station

An area outside Leipzig, Germany, has one of the world's largest solar-powered electricity generators. A field of 33,500 solar panels directly captures the sun's heat and converts it to electricity. The station is connected to the area's **power grid** and now generates enough electricity to power 1,800 homes.

A solar house

Individuals can also set up their own solar energy system. This involves having solar collectors connected to a system of batteries that store the energy collected when the sun is shining. These systems are becoming very popular. For example, more than 20,000 small-scale solar-power generator systems are purchased each year in Kenya.

Fast fact
Setting up solar energy systems still costs more than using electricity generated by fossil fuels.

Running out of wood

Since the earliest times, humans have used wood as fuel. Burning traditional fuels, such as wood, is still a major source of energy in developing countries, particularly in Africa. Today, however, traditional fuels are not being replaced as fast as they are being used, causing major problems, particularly in developing countries.

Renewable biomass fuels

Biomass fuel comes from living things, such as plants. Biomass is a renewable energy source because it can be grown to provide continuous supplies. Wastes from food crops can be used as biomass fuels.

Limits to grown fuel

Traditional biomass fuel, such as wood, is only likely to play a very small role in energy supplies in industrial countries. However, biomass is an important energy source that people living in rural areas with no other fuels can provide for themselves. Traditional firewood collection is considered unsustainable or nonrenewable if used too intensively.

In theory, biomass fuel such as wood is a renewable source of energy.

The collection of firewood in Africa consumes huge amounts of time and human energy.

CASE STUDY
Firewood in Africa

The main fuels used in Africa include wood, animal dung, leaves, and grasses. Firewood collection along with **deforestation** by logging is removing large amounts of plant cover in Africa.

Fuel of the poor

Many Africans, particularly rural women, spend hours each day searching for enough fuel to be able to cook their evening meals. African families living in cities may spend half of their whole income on fuel for cooking. For poorer African nations, trees often supply 70 to 90 percent of energy.

Environmental damage

Collecting firewood faster than it grows strips the land of vegetation. Firewood collection in much of Africa is at least one-third higher than replacement by regrowth. Removing plant cover then leads to **desertification** of the land, destroying what was useful land. This degrading of land is particularly bad in arid areas when there is high growth in population.

Fast fact
Africa has very low levels of electricity consumption per person. Most Africans use 60 to 200 times less energy than most Europeans.

Toward a sustainable future: New biomass fuels

A sustainable future involves developing a range of renewable fuel sources, which can include biomass fuels. Currently, much biomass fuel use is not sustainable, but alternative approaches are being developed. Growing lumber and other biomass fuels is becoming increasingly important to ongoing fuel supplies.

Modern biomass technologies

Modern biomass technologies are developing fuels that are as easy to use as oil, but are grown by plants. They should also have the advantage of burning more cleanly and not producing the pollution associated with fossil fuels.

In a few places, biomass fuels are a major source of energy. In Brazil, sugar cane is made into a fuel called ethanol. Ethanol provides about half the fuel for transportation in Brazil. In China, gas for fuel is being made from animal dung.

Recycled waste

Organic waste and water plants can also be used to produce **methane** or "biogas." Research is continuing to find ways to generate biomass energy using wastes.

Ethanol is as easy to use for fuel as gasoline.

E-85 ETHANOL

$ 0 0 . 0
TOTAL SALE

MEETS MICH. QUALITY & PURITY STANDARDS

E 85
85% Ethanol

0 0 0 . 0
GALLONS

2 7 9 9
$ PRICE PER GALLON INCLUDING TAX

Oil from this crop of rapeseed will be made into biomass fuel.

CASE STUDY
Biomass fuel farms

Research is being done to find plants that can be grown to produce biomass fuel for transportation.

Land for biomass fuels

One major limitation of biomass energy is the land needed to grow it. One estimate suggests that to grow fuel for all the cars in the world we would have to double the amount of land in use today for farming. Research is being done on developing biomass fuel farms in areas that are unable to grow food crops. Switchgrass, for example, is a North American desert plant that survives where little else can grow.

Algae biomass fuel

Research is progressing into growing microscopic **algae** in plastic tubes to provide biomass fuel. Some green algae produce **hydrogen** as a waste product when they are in sunlight. If this hydrogen could be collected, it may be useful in hydrogen-powered cars. Algae hydrogen farms could provide a rich source of fuel in the future.

A number of problems still exist with this "hydrogen farming" technology. Researchers hope to develop technology to provide hydrogen fuel as a renewable clean fuel in the near future.

Problems with energy alternatives

Each of the energy alternatives that have been developed have strengths and weaknesses. The disadvantages of nonrenewables, such as nuclear and geothermal energy, are well known. However, renewable alternatives, including wind and solar power, also have disadvantages.

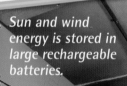

Sun and wind energy is stored in large rechargeable batteries.

Disadvantages of nonrenewable alternatives

Nuclear energy

Mining uranium can cause pollution.

Nuclear waste is very dangerous and must be sealed up and buried for many years.

Accidents at nuclear power plants can be major disasters.

Geothermal energy

Not many places are suitable for geothermal power stations.

Poisonous gases and minerals may come up from underground.

Sometimes a geothermal site may "run out of steam."

Disadvantages of renewable alternatives

Wind energy

Huge wind turbines create visual and noise problems and can also kill birds.

Energy needs to be able to be stored for use when the wind is not blowing.

It is not a good alternative for transportation.

Solar energy

It can be costly to set up.

Energy needs to be able to be stored for use when the sun is not shining.

At present, it is not a good alternative for transportation.

Nuclear power plants do not release greenhouse gases, but there are other problems with wastes.

CASE STUDY

Nuclear power as an alternative fuel source

In 2006, there were about 440 nuclear reactors across the globe, producing about 16 percent of the world's electricity. Several hundred more reactors are due to be built by 2030. Nuclear power already provides most of the electricity in France.

The advantages of nuclear power

Nuclear power plants produce huge amounts of energy for electricity. **Radioactive** material produces much more energy than the same amount of fossil fuel.

Nuclear power plants do not release carbon dioxide or add to global warming. Well-constructed power plants do not release radioactivity into the **atmosphere**.

The disadvantages of nuclear power

Mining uranium is environmentally damaging, and supplies will run out in about 50 years.

Safety is a major problem, and an accident at a nuclear power plant can cause major health and environmental problems.

Radioactive waste from the power plants is toxic to living things for 710,000 years, and there is no safe way to store or dispose of it.

Fast fact
Construction of nuclear power plants declined due to safety fears following the 1979 Three Mile Island accident in the U.S., and the 1986 disaster at Chernobyl, Ukraine.

Toward a sustainable future: Clean, plentiful energy

For a sustainable future, clean, nonpolluting, and renewable energy sources are urgently needed.

Choosing between alternatives

It is important that people understand the strengths and weaknesses of each alternative energy source being developed. These need to be considered when choosing which energy sources to use. Just as fossil fuels have created environmental problems including air pollution and global warming, alternative fuels can also create damage to the environment.

Using a range of fuels

In the future, most areas will be supplied by a combination of energy sources. Alternatives such as wind, solar, and biomass energy are being used in addition to traditional sources of energy in different areas. It is predicted, for instance, that the use of natural gas to generate electricity will increase worldwide from 18 percent in 2002 to 24 percent in 2025.

Fast fact
Zero-energy buildings produce as much or more energy than they use by combining energy-saving construction and appliances with solar and wind technologies.

This zero-energy home in the U.S. uses passive solar design to reduce energy requirements for heating to a small fraction of the requirements of an average home.

World energy resources 2001

Traditional biomass 10 %
Large hydro 5.7 %
Nuclear 3.8 %
"New" renewables 2.3 %
Oil 34.1 %
Natural gas 21.6 %
Coal 22.5 %

In 2001, about 78 percent of the world's energy was from fossil fuels.

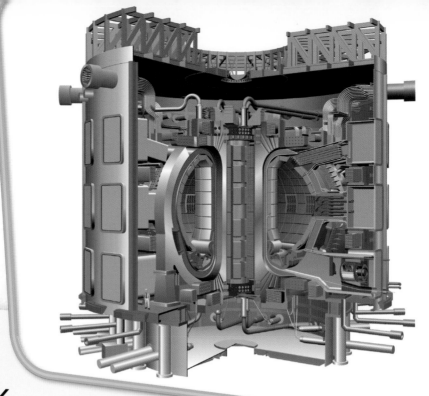

A fusion power design.

CASE STUDY

Fusion—the wonder energy source

One completely untapped source of energy being investigated is fusion power. In the future, fusion research may provide a widely available, unlimited energy source with low environmental impacts.

What is fusion power?

Fusion is the energy source in the sun. Fusion power is produced when two atoms fuse into a larger atom. This happens at very high temperatures when hydrogen turns into a special type of gas, called plasma.

The machine used to capture the energy of fusion is called a Tokamak. Spherical Tokamaks or STs are being tested in laboratories including the ROTAMAK ST in South Australia, SPHERA in Italy, and MAST and START in the U.K.

Project ITER

In June 2005, a hundred-acre (40 ha) site in France was selected for the first experimental large-scale fusion power station, called Project ITER. A number of countries are working together on the project. Project ITER will test the theory that fusion can be used on Earth. The power station is expected to begin operation in 2016.

Fast fact
ITER is pronounced as in the word "litter" and means "the way" in Latin.

What can you do?
Save energy

You may think that just one person cannot do much, but everyone can help. If every person is careful, the little differences can add up.

Use less energy at home

You can help save energy by reducing your energy consumption. You could conduct a home energy audit to find how you might save on energy. An energy audit locates wasteful energy use, including heating and cooling, lighting, and electrical appliances.

Here's what to look for:

- Check for air leaks and drafts such as gaps along the edges of flooring, walls, doors, windows, and ceilings.

- Check the lightbulbs in your house. Do you have low-energy lightbulbs? You may have brighter bulbs where lower energy ones would do. Keep a record of when you use lights to see if they are switched off when not needed.

- Check whether any electrical appliances use energy in a standby mode when not in use. Switch them off at the electrical outlet when not in use.

Fast fact
A 20-watt compact fluorescent bulb will save about 1,430 pounds (650 kilograms) of greenhouse gas over five years, compared with using standard incandescent 100-watt lightbulbs.

*It takes less energy to light up your home with compact fluorescent lights than with traditional **incandescent** lighting.*

Use energy-efficient lighting

Lighting consumes about 15 percent of a household's electricity use. Energy-efficient lighting technologies can reduce the amount of energy used to light homes. Fluorescent lights, including compact fluorescents, generate one-fifth of the greenhouse gases that ordinary lightbulbs produce.

Incandescent lighting

Traditional incandescent electric lighting is the most common type of lighting used in homes. Incandescent lamps have a low efficiency and a short operating life of about 750 to 2,500 hours. Incandescent lamps are less expensive to buy. However, because they use more power and have short life spans, they are more expensive to operate.

Low-energy lights

Low-energy lights such as compact fluorescent lights give out more light for the energy they use and less heat than incandescent lamps. They can save 75 to 90 percent of lighting energy when they replace incandescent lights.

Toward a sustainable future

Well, I hope you now see that if you accept my challenge your world will be a better place. There are many ways to work toward a sustainable future. Imagine it . . . a world with:

- a stable climate
- clean air and water
- nonpolluting, renewable fuel supplies
- plenty of food
- resources for everyone
- healthy natural environments

This is what you can achieve if you work together with my natural systems.

We must work together to live sustainably. That will mean a better environment and a better life for all living things on Earth, now and in the future.

Web sites

For further information on energy, visit these Web sites:

- Electronic Green Journal http://egj.lib.uidaho.edu/index.html
- World Energy Council http://www.worldenergy.org/wec-geis/edc/
- Fusion power http://www.iter.org/index.htm

Glossary

acid rain
rain containing acids which falls from polluted skies

algae
living things that are found in water and make food using the energy from the sun

atmosphere
the layer of gases surrounding Earth

carbon dioxide
a colorless, odorless gas in the atmosphere

climate change
changes to the usual weather patterns in an area

deforestation
removal or clearing of forest cover

degraded
run down or reduced to a lower quality

desertification
turning an area into desert, with low plant cover and a high risk of erosion

developed countries
countries with industrial development, a strong economy, and a high standard of living

developing countries
countries with less developed industry, a poor economy, and a lower standard of living

fossil fuels
fuels such as oil, coal, and gas, which formed underground from the remains of animals and plants that lived millions of years ago

geothermal
heat from inside Earth

global warming
an increase in the average temperature on Earth

greenhouse gases
gases that help trap the sun's heat in the atmosphere

habitats
areas used by living things to provide their needs

hydroelectricity
power generated by moving water in rivers and dams

hydrogen
a gas that can be burned as fuel and which produces only water

incandescent
giving off light at a high temperature

lubricants
substances used to decrease rubbing or friction between moving surfaces

methane
a gas that is given off from burning fossil fuels and decomposing vegetation (including the digestion of plants by animals)

nonrenewable
a resource that is limited in supply and which cannot be replaced once it runs out

power grid
system of wires to carry electricity

radioactive
material that produces waves of energy, called radiation

renewable
a resource that can be constantly supplied and which does not run out

sustainable
a way of living that does not use up natural resources

United Nations Environment Program
a program, which is part of the United Nations, set up to encourage nations to care for the environment

Index